FLORA OF TROPICAL EAST AFRICA

JUNCACEAE

Susan Carter

Annual or perennial herbs, rarely shrubs. Roots fibrous. Stems erect, cylindric or rarely compressed, naked or leafy, sometimes rhizomatous or stoloniferous, and then horizontal or ascending. Leaves grass-like or cylindric, sheathing at the base, sometimes reduced to cataphylls; sheaths open and sometimes auriculate, or closed. Inflorescence terminal, often pseudolateral, compound or rarely simple with one flower, umbellate, paniculate, the flowers solitary or in capitula (heads); bracts leaf-like, scarious or membranous, persistent; bracteoles sometimes present. Flowers small, regular, ♀; rarely the plants dioecious. Perianth-segments 6, in two series, subequal, glumaceous, green or brown, usually membranous at the edges. Stamens 6, opposite and shorter than the perianth-segments, the 3 inner sometimes absent; filaments linear or triangular; anthers basifixed, 2-thecous, introrse, dehiscing longitudinally. Carpels 3, joined; ovary superior, 1- or 3-locular; style rarely absent; stigmas 3. Fruit a loculicidal capsule. Seeds 3 or many, ovoid to obovoid, apex sometimes apiculate, base sometimes tailed.

A family of 8 genera of world-wide distribution concentrated in temperate zones.

A comprehensive account of the synonymy can be found for most cases in Buchenau, E.P. IV, 36 (1906).

Leaves and bracts glabrous throughout; seeds many . . 1. **Juncus**
Leaves and bracts with long hairs on the margins; seeds 3 . 2. **Luzula**

1. JUNCUS

L., Sp. Pl.: 325 (1753) & Gen. Pl., ed. 5: 152 (1754); Buchen. in E.P. IV, 36: 98 (1906); Weim. in Svensk Bot. Tidskr. 40: 141 (1946)

Annual or perennial glabrous, tufted or rhizomatous herbs. Leaves grass-like and flat or channelled, or cylindric and smooth, sometimes septate, sheathing below, sometimes reduced to cataphylls; sheaths open, rarely closed. Inflorescence compound, panicled, bracteate; bracteoles sometimes present, placed immediately below and clasping the flower. Flowers ♀. Perianth-segments free, ovate or lanceolate, entire, persistent. Stamens 3 or 6, attached at the base of the segments; filaments filiform or flattened and triangular; anthers linear or oblong. Ovary sessile; style filiform; stigmas covered with elongated, sticky papillae. Capsule 1- or 3-locular, or incompletely 3-septate. Seeds many.

A large genus of world-wide distribution, principally of temperate zones, but occurring also in arctic regions and on mountains and highlands in the tropics; usually in damp situations, sometimes in maritime habitats.

Attempts have been made in the past to separate *Juncus* into a number of genera, none of which have been successfully maintained. However, Buchenau in E.P. IV, 36:

99 (1906) divided it into 8 subgenera which have since been upheld by most authors. Of these the following 4 are represented in East Africa: subgenus II, *Junci poiophylli* Buchen. (sp. 1); III, *Junci genuini* Buchen. (sp. 2); V, *Junci septati* Buchen. (sp. 3); VIII, *Junci graminifolii* Buchen. (spp. 4, 5). The names themselves are not, however, strictly in accordance with the present 'International Code of Botanical Nomenclature'.

Annuals and perennials; each flower subtended by a bract
   and 2 bracteoles:
   Annual; stem leafy; no sheathing cataphylls evident
      at the stem-base .   .   .   .   .   .   1. *J. bufonius*
   Perennial; stem naked; sheathing cataphylls present
      at the stem-base .   .   .   .   .   .   2. *J. effusus*
Perennials; each flower subtended by a single bract:
   Leaves cylindric and septate .   .   .   .   3. *J. oxycarpus*
   Leaves flat and grass-like:
      Leaves less than 5 mm. wide at the base; inflorescence
         contracted, with the longest branches not more
         than 3 cm. long, and with not more than 8
         capitula .   .   .   .   4. *J. dregeanus*
                                           subsp. *bachitii*

      Leaves more than 5 mm. wide at the base; inflores-
         cence lax, with the longest branches at least
         6 cm. long, and with at least 10 capitula .   5. *J. engleri*

1. **J. bufonius** *L.*, Sp. Pl.: 328 (1753); F.T.A. 8: 95 (1901); Buchen. in E.P. IV, 36: 105 (1906). Type: Europe (LINN, lecto.)

Small tufted annual up to 20 cm. high. Cataphylls absent or usually one to each stem, membranous, 1–1·5 cm. long. Leaves basal and cauline, 3–6 to a stem, grass-like, slightly channelled above, 3–15 cm. long, 1–2 mm. wide; sheaths open, 0·5–1·5 cm. long, with membranous margins. Inflorescence taller than the leaves, many-branched, consisting of 1-sided cymes with branches up to 10 cm. long, and flowers laxly arranged and solitary, occasionally bunched together in groups of 2–6; lower bracts 1–4, leaf-like, up to 8 cm. long; upper bracts membranous, about 5 mm. long; flowers on 0·5–2 mm. long pedicels; each flower subtended by a bract and 2 bracteoles; bracts membranous, subtriangular, 3–4 mm. long; bracteoles membranous, subtriangular, 1·5–2 mm. long. Perianth-segments lanceolate, acute, green with membranous margins, becoming pale brown and rigid at the fruiting stage; outer segments 5–6 mm. long, 1·25 mm. wide; inner segments 4·5–5 mm. long, 1 mm. wide. Stamens 3 or 6; filaments linear, 1·25 mm. long; anthers linear, 1 mm. long. Ovary 3-locular; style 0·5 mm. long; stigmas 1 mm. long, strongly papillose. Capsule trigonous, obovoid, truncate and apiculate at the apex, shiny red-brown. Seeds ovoid, about 0·4 mm. long, red-brown.

KENYA. Nakuru District: Ol Joro Orok, Dec. 1956, *G. R. van Someren* 543!; Naivasha District: Kipipiri, Mananga, 16 Feb. 1956, *Sherwood* 36!
DISTR. **K3**; North Africa, Ethiopia and South Africa (Cape Province) and throughout the world in temperate regions and mountains and highlands of the tropics
HAB. Waterlogged soil, or damp places near streams in upland grassland and cultivated land; 2400–2750 m.

VARIATION. A very variable species for which several varieties have been described depending upon the laxity or tightness of the inflorescence.

2. **J. effusus** *L.*, Sp. Pl.: 326 (1753); F.T.A. 8: 92 (1901); Buchen. in E.P. IV, 36: 135 (1906); Weim. in Svensk Bot. Tidskr. 40: 143 (1946). Type: Europe (LINN, lecto.)

Perennial densely tufted erect herb to 120 cm. high. Rhizome short, internodes scarcely 5 mm. long. Stems to 5 mm. in diameter, leafless, green, finely striated, the sterile ones sharply pointed. Leaves reduced to 3–6 cataphylls, 1–20 cm. long, straw-coloured to dark purple-brown, with the longer inner 1 or 2 bearing a reduced filiform cylindric deciduous leaf-blade up to 1 cm. long. Inflorescence pseudolateral, many-branched, consisting of 1- and 2-sided cymes, dense or lax, 1·5–8 cm. in diameter, subtended by a cylindric sharply pointed bract up to 30 cm. long, which resembles a continuation of the stem; branches many, from a few mm. up to 5 cm. long, each with a scarious lanceolate bract 2–5 mm. long at its base; flowers on slender 0·5–5 cm. long pedicels; each flower subtended by a bract and 2 bracteoles; bracts triangular, 1–1·5 mm. long, acute, membranous, buff-coloured; bracteoles similar, 1 mm. long. Perianth-segments lanceolate, subulate at their apices, green with colourless to brown membranous margins; outer segments 1·75–2·5 mm. long, 0·5 mm. wide; inner segments slightly shorter. Stamens 3; filaments linear, 0·5 mm. long; anthers linear, 0·4–0·5 mm. long. Ovary 3-locular; style very short; stigmas 0·75 mm. long. Capsule trigonous, obovoid, 2–2·5 mm. long, shortly apiculate, apex slightly depressed at maturity, shiny, olive to red-brown. Seeds ovoid, about 0·4 mm. long, very shortly apiculate, faintly reticulated, red-brown.

UGANDA. Toro District: Ruwenzori, Mobuku Valley, 11 Apr. 1948, *Hedberg* 766 !, Kigezi District: 2 km. from Kabale off Mbarara road, R. Nyakizumba, 9 Feb. 1953, *Norman* 204 !
KENYA. Trans-Nzoia District: Elgon, E. slope above Japata Estate, 1 Mar. 1948 *Hedberg* 180 !; Naivasha District: Kipipiri, towards North Kinangop, 31 Mar. 1957 *Verdcourt* 1771 !; Meru District: NE. Mt. Kenya, Marimba Forest, 14 Oct. 1960, *Polhill & Verdcourt* 313 !
TANGANYIKA. Ufipa District: Mbisi Forest, 9 Nov. 1956, *Richards* 6943 !; Morogoro District: Uluguru Mts., Bunduki, by Mgeta R., 20 Mar. 1953, *Drummond & Hemsley* 1729 !
DISTR. U2; K3, 4; T2, 4, 6; worldwide in temperate regions and mountains and highlands of the tropics
HAB. In swamps and by streams in upland rain-forest, upland evergreen bushland and grassland; 1350–3100 m.

SYN. *J. oehleri* Graebn. in E.J. 48: 506 (1912). Type: Tanganyika, Masai/Mbulu District, Lake Ossirwa, *Oehler* in *Jaeger* 499 (B, holo.†)
J. laxus Robyns & Tourn. in B.J.B.B. 25: 252 (1955). Type: Congo Republic, Kivu Province, Kundhuru-Ya-Tshuve, *de Witte* 1976 (BR, holo. !)

VARIATION. Another very variable species for which many varieties have been described depending upon the laxity of the inflorescence and exact shape of the capsule.

3. **J. oxycarpus** *Kunth*, Enum. Pl. 3: 336 (1841); F.T.A. 8: 93 (1901); Buchen. in E.P. IV, 36: 196 (1906); Weim. in Svensk Bot. Tidskr. 40: 166 (1946). Types: South Africa, Cape Province, Liesbek R., *Bergius* (B, syn.†) & Berg R. near Paarl, *Drège* (K, isosyn !); also other specimens of Drège, not seen.

Perennial tufted herb to 70 cm. high. Stems usually erect, leafy, sometimes trailing and then rooting and branching at the nodes. Cataphylls 1–2, up to 5 cm. long, often tinged reddish, with membranous margins; apex sometimes crowned with a rudimentary leaf. Leaves 2–5 to a stem, cylindrical, septate, up to 25 cm. long; sheaths open, up to 7·5 cm. long, biauriculate, margins membranous; auricles 1–2·5 mm. long. Inflorescence taller than the leaves, branched, consisting of up to 20 subspherical capitula each with 20 or more flowers; branching sometimes condensed and the inflorescence appearing to consist of only 2–3 large capitula; branches from a few mm. up to 8 cm. long in each inflorescence; lower bracts leaf-like, 2–5 cm. long; upper bracts lanceolate, about 5 mm. long, apiculate, scarious; flowers subsessile, each subtended by a single bract; bracts triangular, 2–2·5 mm. long, acute, membranous. Perianth-segments equal, lanceolate, 3·5–4·5 mm. long, 1·25 mm. wide, acute, margins membranous, green when young, becoming reddish-brown. Stamens 3, sometimes 6; filaments linear, 1·25 mm. long; anthers linear, 0·75 mm. long. Ovary 1-celled; style very short; stigmas 1·25 mm. long. Capsule trigonous, oblong, 2·5–3·5 mm. long, apex drawn out and apiculate, shiny, buff-coloured below, red-brown to almost black above. Seeds ovoid, 0·5 mm. long, apiculate, reticulated, red-brown. Fig. 1/8–10.

KENYA. Elgeyo District: Cherangani Hills, forest below Kaisungor, 1 Oct. 1959, *Verdcourt* 2435!; Naivasha District: halfway down Kedong escarpment, 8 Dec. 1954, *Verdcourt* 1161!; Embu District: Mt. Kenya, E. side near Forest Station, 26 Dec. 1921, *Fries* 331!

TANGANYIKA. Lushoto District: W. Usambara Mts., Mtai–Mlalo road near Kidologwai, 19 May 1953, *Drummond & Hemsley* 2644!; Ufipa District: Mbisi Forest, 9 Nov. 1956, *Richards* 6948!; Iringa District: Iheme, 12 Oct. 1936, *McGregor* 5!

DISTR. **K**3, 4; **T**2–5, 7, 8; widespread in Africa from Eritrea, Somali Republic and Ethiopia, through to the eastern Congo Republic, Zambia, Malawi, Rhodesia, Angola and South Africa

HAB. In swamps and marshes and the edges of streams in upland rain-forest, riverine forest and upland grassland; 1400–2750 m.

SYN. *J. quartinianus* A. Rich., Tent. Fl. Abyss. 2: 339 (1851). Type: Ethiopia, Shire, *Quartin Dillon* (P, holo.)
  [*J. fontanesii* sensu Engl., Hochgebirgsfl. Trop. Afr.: 158 (1892) & in Abh. Akad. Wiss. Berlin 1894: 59 (1894), *non* Laharpe]
  J. suboxycarpus Adamson in J.L.S. 50: 14 (1935). Type: South Africa, Natal, *Schlechter* 3043 (K, iso.!)
  J. oxycarpus Kunth subsp. *sparganioïdes* Weim. in Svensk Bot. Tidskr. 40: 166 (1946). Type: Mt. Kenya, *Fries* 1477 (UPS, holo.!)

NOTE. This species, confined to the African continent, is very closely allied to *J. fontanesii* Laharpe from the Mediterranean region, a smaller, more slender plant with fewer flowers in its capitula and long-beaked capsules. Some tropical African specimens of *J. oxycarpus*, notably from Kenya, Ethiopia and Eritrea, differ from the majority of South African specimens by having slightly larger flowers. This form was described as subsp. *sparganioïdes* by Weimarck. However, the extensive material examined from the countries cited above shows the existence of a complete range of intermediate forms, so the subspecies is here reduced to synonymy.

4. **J. dregeanus** *Kunth*, Enum. Pl. 3: 344 (1841); Buchen. in E.P. IV, 36: 251 (1906); Weim. in Svensk Bot. Tidskr. 40: 158 (1946). Type: South Africa, " between Cape Colony and Port Natal ", *Drège* 4387 (B, holo.†)

Perennial, densely tufted herb to 45 cm. high. Stems erect, leafy at the base. Cataphylls absent. Leaves linear, up to 25 cm. long, not more than 5 mm. wide at the base, mucronate, slightly channelled; sheaths up to 4 cm. long, margins wide, membranous, closed for up to half their length, becoming split, tinged red. Peduncle leafless. Inflorescence taller than the leaves, branched, consisting of up to 8 subspherical capitula each with about 6–10 flowers; branches never more than 3 cm. long, usually contracted and the inflorescence appearing to consist of 1–3 capitula; lower bracts leaf-like, up to 6 cm. long, with the lowest usually overtopping the inflorescence; upper bracts lanceolate, 3–4 mm. long, scabrous; flowers subsessile, each subtended by a single bract; bracts triangular, 3·5 mm. long, 1·5 mm. wide, membranous. Outer perianth-segments lanceolate, 3 mm. long, 1 mm. wide, mucronate; inner segments lanceolate, 2·75 mm. long, 1 mm. wide, acute, margins membranous, colourless. Stamens 3–6; filaments linear, 0·75 mm. long; anthers linear, 0·5 mm. long. Ovary 3-locular; style 0·2 mm. long; stigmas 1 mm. long. Capsule subglobose, 2–2·5 mm. long, 1·5–2 mm. in diameter, apiculate. Seeds ovoid, 0·23–0·53 mm. long, apiculate.

subsp. **bachitii** (*Steud.*) *Hedb.*, Afroalp. Vasc. Pl.: 61, 263 (1957). Type: Ethiopia, Tigré Province, Mt. Bachit, *Schimper* 114 (B, holo.†)

Flower-bracts brown. Perianth-segments entirely blackish-brown, except for the membranous margins of the inner segments. Stamens 6. Capsule 2·25–2·5 mm. long, 1·75–2 mm. in diameter. Seeds 0·37–0·53 mm. long (*fide* Hedberg). Fig. 1/1–7.

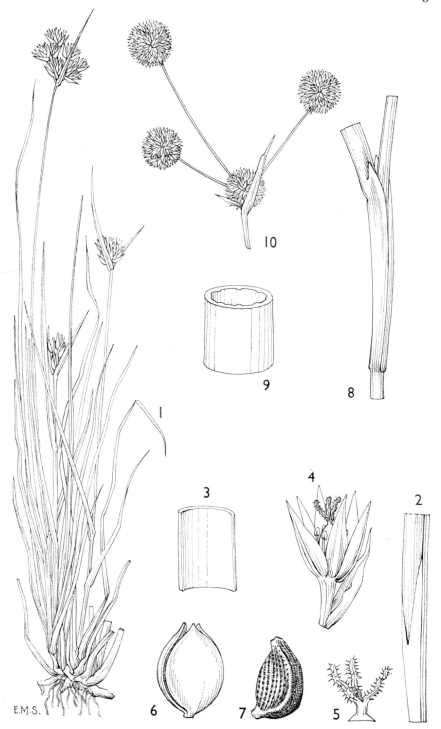

FIG. 1. *JUNCUS DREGEANUS* subsp. *BACHITII*—**1**, habit, × ⅔; **2**, closed sheath, × 6; **3**, section of leaf, × 12; **4**, flower and bract, × 6; **5**, style and stigmas, × 12; **6**, capsule, × 12; **7**, seed, × 60. *J. OXYCARPUS*—**8**, open sheath, × 1½; **9**, section of leaf, × 12; **10**, inflorescence, × 1. 1, 4, 5, from *Fries* 1189; 2, 3, 6, 7, from *Brenan & Greenway* 8218; 8–10, from *Richards* 6948.

UGANDA. Toro District: Ruwenzori, Mobuku Valley, 11 Apr. 1948, *Hedberg* 752 !;
Kigezi District: Muhavura/Mgahinga saddle, Sept. 1946, *Purseglove* 2205 !; Mbale
District: Bugishu, Bulambuli, 4 Sept. 1932, *A. S. Thomas* 573 !
KENYA. Elgeyo District: E. of Kitale, Cherangani Hills, 11 Dec. 1959, *Bogdan* 4991 !;
N. Nyeri District: Mt. Kenya, NW. slopes, 21 Aug. 1948, *Hedberg* 2005 !; S. Nyeri
District: Aberdare Range, near W. part of Nyeri track, 16 July 1948, *Hedberg* 1599 !
TANGANYIKA. Moshi District: S. Kilimanjaro, 11 Mar. 1934, *Schlieben* 4922 !; Ufipa
District: Sumbawanga, Nsanga area, Molo, 1 Jan. 1962, *Robinson* 4858A !; Rungwe
District: NE. of Rungwe Mt., near the upper Kiwira R., 25 Oct. 1947, *Brenan &
Greenway* 8218 !
DISTR. **U**2, 3; **K**3, 4; **T**2, 4, 7; tropical Africa except the west, from the Sudan Republic
and Ethiopia to the Congo Republic, Zambia, Rhodesia and Malawi
HAB. In swamps and in shallow, still water at the edges of streams in upland rain-
forest, upland evergreen bushland, upland and moor grassland; 2100–3400 m.

SYN. *J. bachitii* Steud., Syn. Pl. Glum. 2: 305 (1855)
[*J. indescriptus* sensu Adamson in J.L.S. 50: 24 (1935) & in Journ. S. Afr. Bot. 3:
165 (1937); Weim. in Svensk Bot. Tidskr. 40: 159 (1946), *non* Steud. These
reports of an extension from the Cape into Tanganyika are all based on one
gathering, *Stolz* 1144 (from Tanganyika, Rungwe District, Mbaka R., Kalagwe)]
[*J. dregeanus* sensu Weim. in Svensk Bot. Tidskr. 40: 158 (1946), *non* Kunth
sensu stricto]

VARIATION. The South African subsp. *dregeanus* has lighter coloured flowers, with buff
flower-bracts, and perianth-segments which are sometimes dark but not entirely
blackish-brown. The stamens number 3, sometimes 6, the capsule is slightly smaller,
and the seeds are distinctly smaller, 0·23–0·37 mm. long (*fide* Hedberg).

5. **J. engleri** *Buchen.* in E.P. IV, 36: 248 (1906); Weim. in Svensk Bot.
Tidskr. 40: 158 (1946). Types: Tanganyika, W. Usambara Mts., *Engler*
1060 & 1409 & *Holst* 334a [?] (all B, syn.†)

Perennial, tufted herb to 60 cm. high. Stems erect and trailing, rooting
at the nodes; internodes less than 1 cm. long. Cataphylls absent. Leaves
all basal, linear-lanceolate, 5–12 mm. wide at the base, up to 40 cm. long, with
subulate tips; sheaths closed, becoming split, up to 2 cm. long, tinged red.
Peduncle leafless. Inflorescence taller than the leaves, many-branched,
consisting of up to 50 semi-spherical capitula each with about 10 flowers;
branches 1–8 cm. long; lower bracts leaf-like, up to 4·5 cm. long; upper
bracts lanceolate, 3–4 mm. long, scabrous, blackish-brown; flowers subsessile, each subtended by a single bract; bracts triangular, 3 mm. long,
1·5 mm. wide at the base, acute, membranous, brown. Outer perianth-
segments lanceolate, 2·75 mm. long, 0·75 mm. wide, apex subulate, blackish-
brown; inner segments triangular, 2·5 mm. long, 1 mm. wide, obtuse,
blackish-brown with a wide colourless membranous margin. Stamens 6;
filaments linear, 0·75 mm. long; anthers linear, 0·75 mm. long. Ovary 3-
locular; style 0·3 mm. long; stigmas 1 mm. long. Capsule subglobose,
1·75 mm. long, 1·25 mm. in diameter, apiculate, shiny, red-brown above,
pale below. Seeds ovoid, 0·3 mm. long, smooth.

TANGANYIKA. Pare District: S. Pare Mts., Mtonto, 4 July 1942, *Greenway* 6548 !;
Lushoto District: W. Usambara Mts., Mkusu Valley between Mkuzi and Kifungilo,
23 Apr. 1953, *Drummond & Hemsley* 2214 !
DISTR. **T**3; confined to the Usambara and Pare Mts.
HAB. Wet boggy places and by streams in upland rain-forest and upland grassland;
1400–1900 m.

SYN. [*J. lomatophyllus* sensu Baker in F.T.A. 8: 94 (1901); Buchen. in E.P. IV, 36:
247 (1906); Vierhapper in E. & P. Pf., ed. 2, 15A: 216 (1930), *non* Spreng.]

NOTE. *J. engleri* is closely related to and has been considered synonymous with the
South African and Rhodesian *J. lomatophyllus* Spreng. which has usually slightly
wider leaves, longer anthers, 1·5 mm. long, and a longer style, 1·5 mm. long, resulting
in a distinctly beaked capsule.

2. **LUZULA**

DC., Fl. Fr. 3: 158 (1805); Buchen. in E.P. IV, 36: 42 (1906), *nom. conserv.*

Juncoïdes Adanson, Fam. 2: 47 (1763)

Perennial, rarely annual herbs, tufted, rhizomatous or stoloniferous. Leaves grass-like, flat, basal and cauline, with long hairs on the margins, sheathing below, sometimes reduced to cataphylls; sheaths always closed. Inflorescence compound, panicled or subspicate, bracteate; bracts hairy on the margins; bracteoles present, placed immediately below and clasping the flower. Flowers ☿. Perianth-segments free, ovate or lanceolate, persistent, margins entire or sometimes lacerated. Stamens 3 or 6, attached at the base of the segments; filaments filiform; anthers oblong or linear. Ovary sessile; style filiform; stigmas filiform. Capsule 1-celled; seeds 3, basal, globose or ovoid, often tailed.

A fairly large genus of world-wide distribution, concentrated in the northern temperate zones but found also in arctic regions and mountains and highlands in the tropics; usually in damp situations.

As with *Juncus*, the subgenera of *Luzula* used by Buchenau in E.P. IV, 36: 43 (1906), seem generally acceptable. Only two are represented here: *L. johnstonii* belongs to subgenus I, *Pterodes* Griseb., while *L. campestris* and *L. abyssinica* belong to subgen. III, *Gymnodes* Griseb., which should now be referred to as subgen. *Luzula*.

Inflorescence lax, flowers solitary, each distinctly
    pedicellate .   .   .   .   .   .   1. *L. johnstonii*
Inflorescence dense, subspicate or the flowers in capitula:
    Flowers in separate capitula arranged ± in an umbel,
      the lower branches distinctly spreading   .   .   2. *L. campestris*
                                                  var. *gracilis*

    Flowers crowded into a lobed spike; if the spicate
      capitula are separated then the branches are all
      erect, not arranged in an umbel   .   .   .   3. *L. abyssinica*

1. **L. johnstonii** *Buchen.* in E.J. 12: 79 (1890); F.T.A. 8: 96 (1901); Buchen. in E.P. IV, 36: 45 (1906). Type: Tanganyika, Kilimanjaro, *H.H. Johnston* 28 (K, holo.!, BM, iso.!)

Perennial herb to 60 cm. high, tufted, stoloniferous. Stolons up to 15 cm. long, with internodes about 1·5 cm. long. Cataphylls 1–4, 1–1·5 cm. long, purplish. Basal leaves up to 25 cm. long, 8 mm. wide, all with blunt, hardened tips; margins, especially of young leaves, sparsely covered with long white hairs; bases purplish; sheaths up to 4 cm. long. Peduncle with 2–4 leaves up to 15 cm. long, 6 mm. wide. Inflorescence taller than the leaves, many-branched, consisting of 1- and 2-sided cymes with branches up to 8 cm. long; lower bracts leaf-like, up to 3 cm. long, 3 mm. wide, margins ciliate; upper bracts about 5 mm. long, scarious, purplish with pale membranous laciniate apices; flowers solitary on pedicels up to 12 mm. long; each flower subtended by a bract and 2 bracteoles; bracts lanceolate, 2–3 mm. long, membranous with laciniate apices; bracteoles triangular, 2 mm. long, 1·5 mm. wide at the base, sharply pointed, becoming laciniate, scarious with a membranous margin, dark reddish-brown. Perianth-segments entire, lanceolate, sharply pointed, dark reddish-brown with pale membranous margins; outer segments 3 mm. long, 1 mm. wide; inner segments 3·25 mm. long, 1 mm. wide. Stamens 6; filaments linear, 1 mm. long; anthers oblong, 0·75 mm. long. Ovary trigono-globose; style 1·5 mm. long; stigmas 2·5 mm. long. Capsule subglobose, beaked, 2 mm. in diameter, purplish-brown, shiny; beak 0·5 mm. long. Seeds ovoid, 1·25–1·5 mm. long, dull purplish-brown, distinctly tailed; tail 0·25 mm. long, yellow.

UGANDA. Toro District: Ruwenzori, Mijusi valley, 30 Mar. 1948, *Hedberg* 600 !; Kigezi District: W. slopes of Mt. Muhavura, 3 Oct. 1948, *Hedberg* 2062 !; Mbale District: Elgon, Jan. 1918, *Dummer* 3548 !

KENYA. Trans-Nzoia District: Elgon above Tweedie's sawmill, 2 Mar. 1948, *Hedberg* 213 !; N. Nyeri District: Mt. Kenya, NW. slopes, 18 Aug. 1948, *Hedberg* 1937 !; S. Nyeri District: Aberdare Range, near W. part of Nyeri track, 14 July 1948, *Hedberg* 1568 !

TANGANYIKA. Moshi District: Kilimanjaro, Shira Plateau, Feb. 1928, *Haarer* 1150 ! & above Marangu, south of Peter's hut, 26 June 1948, *Hedberg* 1365 !

DISTR. **U**2, 3; **K**3, 4; **T**2; also in Ethiopia, and the Congo Republic side of the Ruwenzori range and Virunga Mts.

HAB. Damp places, usually in shade in higher part of upland rain-forest and in upland moor; 2400–4200 m.

SYN. [*L. forsteri* forma sensu Oliv. in Trans. Linn. Soc., ser. 2, 2: 352 (1887), *non* (Sm.) DC. A provisional determination of *Johnston* 28 (type of *L. johnstonii*)]
Juncoïdes johnstonii (Buchen.) O. Ktze. (as " *Juncodes* "), Rev. Gen. 2: 724 (1891).

NOTE. This species is closely related to the European and North African *L. forsteri* (Sm.) DC., which is a smaller more slender plant, with narrower shorter leaves and no stolons.

2. **L. campestris** (*L.*) *DC.*, Fl. Fr. 3: 161 (1805); Buchen. in E.P. IV, 36: 83 (1906). Type: Europe (LINN, lecto.)

Tufted perennial up to 60 cm. high; stolons sometimes present. Cataphylls present. Leaves up to 30 cm. long and 6 mm. wide; basal leaves with blunt hardened tips; cauline leaves with acute tips, or blunt and hardened; margins hairy. Peduncle leafy. Inflorescence taller than the leaves, compound, branched; flowers in capitula; lower bracts leaf-like, margins hairy; upper bracts scarious, usually not much laciniate; each flower subtended by a bract and 2 bracteoles. Perianth-segments lanceolate, 2–3 mm. long, pale buff to blackish-purple. Stamens 6. Capsule ± ovoid, mucronate. Seeds 1·2–2·5 mm. long, tailed.

SYN. *Juncus campestris* L., Sp. Pl.: 329 (1753)

var. **gracilis** *S. Carter* in K.B. 17: 179 (1963). Type: Uganda, Elgon, *Dummer* 3545 (K, holo. !)

Loosely tufted herb to 40 cm. high, stoloniferous; stolons few, short, up to 5 cm. long, with internodes about 1 cm. long. Cataphylls 1–2, about 1 cm. long, brown and scaly. Leaves up to 20 cm. long, 5 mm. wide, all with blunt, hardened tips; margins with long white hairs, ± dense on young leaves; sheaths up to 3·5 cm. long. Peduncle with 2 leaves up to 15 cm. long. Inflorescence much taller than the leaves, consisting of up to 8 spreading branches arranged in an umbel, seldom branched above; branches slender, flexuose, up to 7 cm. long, each bearing an ovoid capitulum of up to 15 flowers; lower bracts up to 5 cm. long; upper bracts lanceolate, about 5 mm. long, margins laciniate, reddish. Flowers subsessile; bracts lanceolate, 2 mm. long, 1 mm. wide at the base, membranous with laciniate margins, reddish; bracteoles triangular, ± 1·5 mm. long, scarious, margins laciniate, purplish. Perianth-segments equal, 2·75 mm. long, 1 mm. wide, blackish-purple, pale and membranous at the apex. Filaments linear, 1·25 mm. long; anthers linear, 0·5 mm. long. Ovary trigono-obovoid; style 0·5 mm. long; stigmas 1·25 mm. long. Capsule obovoid, 2 mm. long, 1·5 mm. in diameter, pale reddish-buff. Seeds subglobose, 0·75 mm. long, dark purplish, minutely tailed. Fig. 2.

UGANDA. Mbale District: Elgon, Jan. 1918, *Dummer* 3545 ! & Apr. 1930, *Liebenberg* 1704 ! & by Sala stream, 22 Mar. 1951, *G. H. S. Wood* 126 !

KENYA. Trans-Nzoia District: Elgon, E. slopes above Japata Estate, 1 Mar. 1948, *Hedberg* 186 !

DISTR. **U**3; **K**3; confined to Elgon

HAB. Wet places and by streams in upland grassland and moor; 2700–3650 m.

SYN. [*L. campestris* var. *mannii* sensu Hedb., Afroalp. Vasc. Pl.: 63 (1957), *non* Buchen.]

NOTE. *L. campestris* (L.) DC. var. *mannii* Buchen. in E.J. 12: 159 (1890), is West African, described from specimens collected on Fernando Po and the Cameroon Mt. Var. *gracilis* differs from it by the branches of its inflorescence being longer, more slender and flexuose, and by the lowest bract not overtopping the inflorescence.

Closely allied to *L. campestris* (L.) DC. is *L. africana* Drège, the only species of *Luzula* occurring in South Africa. It has few capitula, and the flowers are distinctly paler than any tropical African species.

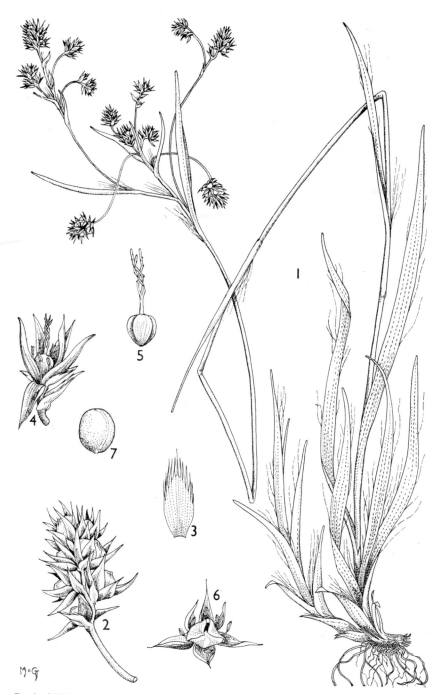

FIG. 2. *LUZULA CAMPESTRIS* var. *GRACILIS*—**1**, habit, × 1; **2**, capitulum, × 4; **3**, flower bract, × 8; **4**, young flower with one perianth-segment pulled down, × 4; **5**, pistil, × 6; **6**, flower with ripe capsule, × 4; **7**, seed, × 12. All from *Dummer* 3545.

3. **L. abyssinica** *Parl.*, Fl. Ital. 2: 310 (1852); Buchen. in E.P. IV, 36: 72 (1906); Hedb., Afroalp. Vasc. Pl.: 62, 264, t. 2/B (1957). Type: Ethiopia, Simen Mts., near Demerki, *Schimper* 1154 (FI, holo., K, iso. !, P, S, iso.—*fide* Hedberg)

Tufted perennial up to 50(–60) cm. high, stoloniferous. Stolons up to 5 cm. long, with internodes about 1 cm. long. Cataphylls 2–3, up to 1·5 cm. long, obtuse, brown and scaly. Leaves up to 30 cm. long (usually about 15 cm. long), 8 mm. wide, all with blunt, hardened tips; margins ± densely covered with long white hairs, especially when the leaves are young; peduncle with (2–)3 leaves up to 20 cm. long; sheaths up to 4·5 cm. long. Inflorescence taller than the leaves, consisting of 2–3 erect branches up to 6 cm. long, bearing up to 10 subsessile contiguous spicate capitula, each of 10–20 flowers, giving the appearance of a single, lobed spike; lower bracts leaf-like, longer than the inflorescence, up to 8 cm. long, 3 mm. wide, with very hairy margins; upper bracts lanceolate, about 5 mm. long, scarious, pale, with densely hairy margins; flowers subsessile, each subtended by a bract and 2 bracteoles; bracts triangular, 2 mm. long, 1·25 mm. wide at the base, membranous, pale, with margins very deeply lacerated; bracteoles about as long as the bracts, with deeply lacerated margins, membranous, reddish. Perianth-segments entire, lanceolate, sharply pointed, dark reddish-brown, sometimes almost black; outer segments 2·5–3 mm. long, 1 mm. wide; inner segments 2–2·5 mm. long, 1 mm. wide. Stamens 6; filaments linear, 1–1·25 mm. long; anthers linear, 0·75–1 mm. long. Ovary trigono-globose; style 0·5 mm. long; stigmas 1·5 mm. long. Capsule shortly apiculate, subglobose, 2 mm. long, 1·5 mm. in diameter, dark reddish-brown above. Seeds ovoid, 0·75 mm. long, very shortly tailed, reddish-brown.

UGANDA. Toro District: Ruwenzori, Bujuku Valley, near Bigo camp, 22 Mar. 1948, *Hedberg* 387 !; Kigezi District: Mt. Muhavura, 19 Oct. 1929, *Snowden* 1563 !; Mbale District: Bugishu, Bulambuli, 4 Sept. 1932, *A. S. Thomas* 537 !

KENYA. Trans-Nzoia District: E. slopes of Elgon, 4 Mar. 1956, *Bogdan* 4139 !; N. Nyeri District: Mt. Kenya, Teleki valley, 31 July 1948, *Hedberg* 1766 !; S. Nyeri District: Aberdare National Park on North Kinangop–Nyeri road, 30 July 1960, *Polhill* 238 !

TANGANYIKA. Moshi District: Kilimanjaro, near Peter's Hut, 23 Feb. 1934, *Greenway* 3769 !; Morogoro District: Uluguru Mts., Lukwangule, 4 Jan. 1934, *Michelmore* 907 !; Rungwe District: Rungwe massif, 21 Jan. 1914, *Stolz* 2442 !

DISTR. U2, 3; K3, 4; T2, 6, 7; also in Ethiopia, and the Congo Republic side of Ruwenzori and Virunga Mts.

HAB. In swamps and damp places from upper parts of upland rain-forest and grassland into upland moor and moor grassland, sheltered or in the open among rocks at higher altitudes; 2000–4550 m.

SYN. *L. spicata* (L.) DC. var. *erecta* E. Mey. in Linnaea 22: 415 (1849). Type: Ethiopia, Simen Mts., near Demerki, *Schimper* 1154 (K, iso. !)
L. macrotricha Steud., Syn. Pl. Glum. 2: 294 (1855), *nom. illegit.* Type: Ethiopia, *Schimper* 1154
L. spicata (L.) DC. var. *simensis* Buchen. in E.J. 12: 128 (1890), *nom. illegit.* Type: Ethiopia, *Schimper* 1154
L. volkensii Buchen, in E.J. 21: 192 (1896). Type: Tanganyika, Kilimanjaro, Mawenzi, *Volkens* 1365 (B, holo.†, BM, K, iso !)
L. abyssinica Parl. var. *volkensii* (Buchen.) Engl. in Urb. & Graebn., Festschr. Asch.: 556 (1904)
L. abyssinica Parl. var. *kilimandscharica* Engl. in Urb. & Graebn., Festschr. Asch.: 556 (1904). Types: Tanganyika, Kilimanjaro, Moshi, *Engler* 1755 (B, syn.†) & above Moshi, *Engler* 1834 (B, syn.†) & above Kibosho, *Uhlig* 1110 (EA, isosyn. !) & Mt. Meru, above Grenze, *Uhlig* 609 (EA, isosyn. !)
L. abyssinica Parl. var. *simensis* (Buchen.) Buchen. in E.P. IV, 36: 72 (1906), *nom. illegit.*

VARIATION. The most frequent and adaptable of the East African *Juncaceae*, this is a variable species, its height, leaf-size and hairiness depending upon the climatic environment (see Hedberg). *L. spicata* (L.) DC. from arctic and alpine regions of the northern hemisphere, to which this species is closely allied, shows the same variation. *L. spicata* differs from *L. abyssinica* Parl. by not having stolons, and by its nodding inflorescence and acutely pointed stem-leaves.

INDEX TO JUNCACEAE

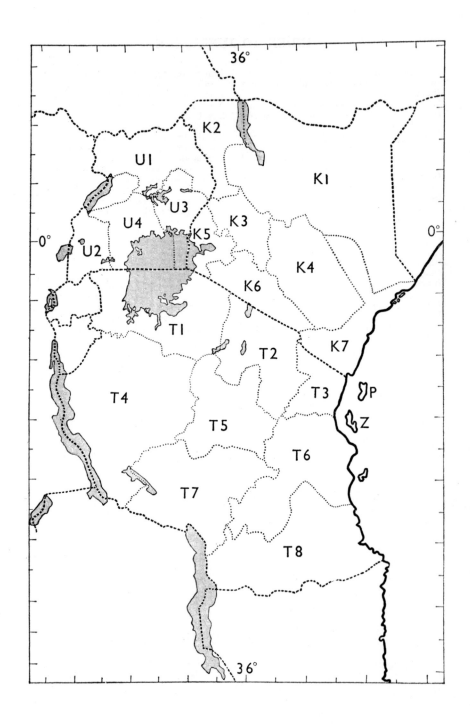